ALIENS
and other
WORLDS

ALIENS
and other
WORLDS

True tales from our solar system and beyond

LISA HARVEY-SMITH
Illustrations by Tracie Grimwood

First published in Australia in 2021
by Thames & Hudson Australia Pty Ltd
11 Central Boulevard, Portside Business Park
Port Melbourne, Victoria 3207
ABN: 72 004 751 964

First published in the United Kingdom in 2022
by Thames & Hudson Ltd
181a High Holborn
London WC1V 7QX

24 23 22 21   5 4 3 2 1

Thames & Hudson Australia wishes to acknowledge that Aboriginal and Torres Strait Islander people are the first storytellers of this nation and the traditional custodians of the land on which we live and work. We acknowledge their continuing culture and pay respect to Elders past, present and future.

ISBN 978-1-760-76116-5 (hardback)

A catalogue record for this book is available from the National Library of Australia

British Library Cataloguing-in-Publication Data
A catalogue record for this book is available from the British Library

Illustration: Tracie Grimwood
Design: Philip Campbell Design
Editing: Rebecca Lim

Printed and bound in China by C&C Offset Printing Co., Ltd

Be the first to know about our new releases,
exclusive content and author events by visiting
thamesandhudson.com.au
thamesandhudson.com
thamesandhudsonusa.com

FSC® is dedicated to the promotion of responsible forest management worldwide. This book is made of material from FSC®-certified forests and other controlled sources.

To Shell, as always

CONTENTS

INTRODUCTION

I found the idea of unidentified flying objects (UFOs) and aliens a bit scary as a child. The only things I knew about life in outer space were from TV shows. You know, the ones with creepy music where people run through dark woods at night, chased by big, goggle-eyed aliens in flying saucers. Now that I'm older, I (thankfully) know that these shows were only make-believe – and that there is so much more to life beyond Earth than what we see on TV.

As I learned more about our universe, I started to think about aliens in a different way. As a scientist, I became fascinated by the planets, stars and galaxies. I read scientific theories of how life on Earth began through the slow transformation of chemicals into living beings. Knowing the science made the possibility of alien life in our universe far less scary and far more intriguing.

Scientists have discovered thousands of worlds beyond our solar system. And science has taught us much about the prospects for life in the universe. Perhaps with further research (and a bit of imagination) we might be able to picture what aliens and their home planets could really be like.

Have you ever tiptoed to the bedroom window at night and peeked through the curtains at a starry sky?

The Moon smiles down with her cool, silver light. As you look around, 1 or 2 bright stars come into view. You'll start to see dozens of tiny sparkles lighting up the night – planets shine in brilliant white or earthy red by the reflected light of the Sun. This is the incredible view of the cosmos from Earth: our home.

More than 7 billion people and trillions of other living things inhabit planet Earth. Huddled together on this humungous chunk of rock, we share our most precious resources: the sunlight, the air and the water. We glide through space at 220 kilometres per second as our world and all its passengers circle the Sun, and the Sun in turn cruises around the centre of our galaxy.

Earth is 1 of 8 planets orbiting the Sun. Can you name them?

Starting closest to the Sun and going further out, we find Mercury, Venus, Earth, Mars, Jupiter, Saturn, Uranus and Neptune. These planets, along with the Sun, join 200 billion other stars in our galaxy. We call our galaxy the Milky Way; it is one of trillions of galaxies making up everything that exists in the universe.

Our goal as astronomers (scientists who study the stars) is to understand our place in the universe, and some of us hunt for evidence of life outside Earth. But in all the vastness of space, we have never discovered another living creature outside our planet. Does life exist out there? Or are we alone in the universe? Right now, we simply do not know.

Scientists and engineers have built rockets, satellites and space probes to search every corner of our solar system for life. We have studied millions of stars with giant telescopes, many of them orbiting Earth in space. We even have teams of astronauts and cosmonauts (that's the Russian word for astronauts) living and working in space to test whether life can survive elsewhere in the galaxy.

As we learn more about our universe, we have more questions about the possibility of finding life beyond Earth.

Will we find other planets that are teeming with life? And if there is life out there, what is it like? Could it take the form of strange birds soaring in the skies of far-away planets, or peculiar plants growing on a distant world? How do aliens eat? Do they play football? Can they live forever? And do aliens have pets?

These are some of the curious questions and mind-blowing mysteries of life in the universe. Let's pull on our spacesuits and go on an interstellar journey (a trip between the stars) to visit the remarkable alien worlds where one day, we might meet our neighbours!

Part 1
Life On EARTH

How did LIFE ON EARTH begin?

Humans are curious creatures. We have always wondered where life came from, and how it all began. That's why we ask, *Which came first, the chicken or the egg?* Although we don't know all the answers, science can help us figure out how this weird old ball of space rock became the haven of life it is today.

Our planet formed more than 4.5 thousand million years ago. More than a billion trillion (1,000,000,000,000,000,000,000) tonnes of dust, gas and rocks were moulded by the force of gravity to create Earth. At the same time, the Sun and planets like Venus, Jupiter and Mars were also growing in the cloud of gas and rubble that made our solar system.

When our Earth was still very young, it had no oceans and no atmosphere. Gradually, gravity scrunched loose piles of rocks and dust together into a tight ball. The pressure of all that rock pulling down in one place squeezed Earth, causing friction that warmed our planet from the inside out. The breakdown of radioactive chemicals like uranium, potassium and thorium heated it further. Meteorites (rocks from space) rained down on our planet, crashing into the surface and turning up the temperature even more. Under these extreme conditions, the young Earth melted, forming a thick and sticky magma.

Heavy metals, like iron and nickel, sank to the middle of the planet and created a dense core. The gooey layer that surrounds the core, called the mantle, became a mess of molten rock that still sloshes and bubbles today, like a boiling pan of water. The lighter materials floated to the top. Gradually, Earth's surface cooled down and formed a solid crust that's up to 10 kilometres thick beneath the oceans and 50 kilometres thick on land.

How do we know what Earth was like before we existed? Using some clever detective work, scientists have figured it out.

Our planet's secrets were hidden in the Jack Hills in Western Australia, where scientists discovered the oldest rocks in the whole world. These 4.4 billion-year-old dirtballs were made by a prehistoric volcano that spilled hot lava into Earth's brand-new oceans. The gases trapped in the rocks tell us a lot about the history of Earth.

By studying ancient rocks from around the world, we have learned that volcanic eruptions were very common in the past. Lava oozed out of the ground like hot chocolate sauce. When it cooled, it became solid, making the land. The smelly gases trapped underground escaped too and created an early version of our atmosphere (although it didn't yet contain the oxygen that we breathe). Steam escaped, which cooled down and became liquid, making ours a watery world. More water arrived in the form of ice on board comets and asteroids that smashed into our planet, slowly filling our oceans.

Around 4 billion years ago, something amazing happened. The first life forms appeared.

All creatures have something in common, whether you are a microbe (a teeny, tiny, living thing, like a bacteria or germ, which is too small to see without a microscope) or a mushroom, a porcupine or a person – we are all made from the same chemical foundations. These include sulphur, phosphorous, oxygen, nitrogen, carbon, and hydrogen. And if you look back far enough, we all share an ancestor; most likely a tiny critter that emerged from a hot steamy bath a few million years after Earth's oceans formed.

The first living thing was probably made when hot volcanic gases escaped from deep inside the planet and mixed with water. This could have happened either on land, in hot volcanic springs, or deep under the ocean, close to vents of hot volcanic gas. Either way, it's likely that the water and gases reacted together like a giant chemistry experiment to make organic compounds, called amino acids. These gradually joined together like the carriages of a train to create the first tiny chemical machines. Once these chemicals were able to make new copies of themselves, life on our planet had truly begun.

But how did we learn all of this from some rocks? Well, some contain the preserved remains (fossils) of ancient microbes that lived billions of years ago, the imprints of which are left permanently in the rock. We call these microbial fossils 'stromatolites' (stro-mat-o-lites). They might be above the ground today, but they may originally have been under, or on the shores of, an ocean.

We know where these microbial fossils once lived because we see the same types of microbes living today in the waters of Shark Bay in Western Australia. Stromatolites are sticky microbial mats that line the seafloor and contain hundreds of thousands of microbes. These mats build up layer by layer over time to form mounds in the shallow water.

The discovery of the oldest known fossils of life on Earth is pretty incredible. Studying them helps us imagine how living things may have grown from non-living things (water and gases) – a process that seems almost magical, but is explained by science.

There is still so much to discover about the beginning of life on Earth, and about life in the universe. For starters, we're a bit more complicated than sticky microbe mats or single-celled life. So, what happened next?

WHY are there 140,000 TYPES of MUSHROOM?

Life on our planet today is made up of more than 1 trillion (1,000,000,000,000) different living things. Earth is teeming with creatures: on land, in the air and even deep within the oceans and Earth's crust. Our planet provides a rich and fertile environment with plenty of food and warming energy from the Sun. It's the perfect place for you, me and all living things.

Life may have existed on our planet for around 4 billion years, but it hasn't always been as varied as it is today.

At first, there were no birds, plants or fish. There were only tiny microscopic creatures, each made up of a single cell.

Cells are little building blocks that make up living things. Each cell contains a special instruction book called DNA (that stands for de-oxy-ribo-nucleic acid, if you're interested), which instructs the cells on what they should look like and how to behave. Some creatures, such as bacteria, are made of just 1 cell. But large creatures, like trees or animals, can each contain billions of cells.

Because each cell contains a design template, our cells can recreate themselves. That's why if you graze your knee, you can grow new skin to make you good as new. It's the same story with your hair, when you get it cut, it simply grows back!

DNA also tells babies what to look like. For example, a baby black-bird will develop wings, feathers, eyes and feet as it grows inside the egg because its parents passed on their genetic design. It is very important that living things are able to copy themselves, otherwise our bodies (and our species) could not survive for long.

Sometimes, when a cell makes a copy of itself, the DNA changes a little bit. We call this a genetic mutation. When the DNA changes, the next generation of cells turn out to be slightly different from the last. Over millions of years, these random genetic mutations have added up to significant changes. This is the basis of the theory of evolution, and it's thanks to evolution that we now have such a dazzling array of plants, animals, fungi and microbes – including more than 140,000 types of mushroom and 34,000 species of fish!

But while some species have thrived as the conditions on the planet have changed over the years, some have died out entirely – becoming what scientists call 'extinct'.

One particularly famous example of this is the dodo. This flight-less bird was common in Mauritius, a small island near Madagascar in the Indian Ocean. When Europeans arrived on the island in the 1500s, they brought with them hungry sailors, as well as monkeys, pigs and rats. As a result, the dodos (which couldn't fly) suddenly found themselves hunted, with no way to fly away or protect themselves. By around 1690, every last dodo was gone.

Just like the dodo, millions of other species have become extinct. In the past, this might have been as a result of competition for food or changes to Earth's climate (such as an ice age). In more recent times, many creatures have died out because of the actions of humans. When we cut down forests, change the course of rivers or pollute our atmos-phere and waterways, other living things struggle to survive.

Extinctions can also happen when the environment changes sud-denly. One such event happened around 66 million years ago, when a

10 kilometre-sized rock from outer space slammed into Earth without warning. It hit the ground with astonishing energy, leaving a gigantic crater in the Gulf of Mexico and throwing up a layer of dust that travelled around the globe. After blocking sunlight and choking the atmosphere for at least a year, this dust settled down to form a thin layer that we still find in rock samples today.

When testing this thin layer of dust within the rocks, scientists have found that it contains a chemical called iridium. This substance is very rare on Earth but is commonly found in asteroids. This detective work by geologists (scientists who specialise in rocks) explains how we know that it was an asteroid and not something else that caused this devastating global catastrophe.

According to fossil records, the asteroid impact resulted in about three quarters of all living creatures to be wiped out. Many species of plants, insects, fish, reptiles and all of the land-based dinosaurs were lost. Interestingly, fungi did well, because they adapted to use the energy from all the dead and rotting leaves and plants to survive.

Next time you look at a mushroom, remember how its tough ancestors managed to outlive the dinosaurs!

Extremophiles

Extremophiles (pronounced extreme-o-files) live up to their name, thriving in the most unlikely places you can imagine.

Have you ever experienced weather that is swelteringly hot, or freezing cold? The temperature in Australia has soared as high as 50 °C in the past! But the hottest I've experienced is 48 °C – on those days I'll be found indoors, in front of a fan, trying to stay cool. The coldest I have ever been was on a very chilly -15 °C in London, England. Even wrapped up in coats, a hat and a scarf, my face was freezing, and it hurt to breathe as the icy cold air entered my lungs.

Very hot and cold weather makes us feel really uncomfortable. That's because our bodies are chemically fine-tuned to work in a specific range of temperatures. Our core body temperature generally lies somewhere between 36.5 °C and 37.4 °C, the perfect environment for the processes that keep us alive. When we get too hot, we sweat – water pools on our skin and evaporates, taking the excess heat with it. If we become too cold, we shiver and our hair stands on end, trying hard to insulate us and take us back to the temperature sweet spot that our body likes.

Our whole environment is important, not just the temperature. We have evolved over millions of years to live on the land, breathe oxygen from the air and drink clean, fresh water from rivers and lakes. We need to experience warm sunshine and eat fresh foods in order to thrive, not bake in the desert heat of California's Death Valley or shiver in darkness at the icy South Pole. In order to adapt to different conditions, we need to use our ingenuity and cunning to survive.

Some humans have developed clever ways to survive for long periods in extreme conditions. Native peoples of northern Canada, Russia and the Arctic live in places where the temperature can plummet down to -50°C. They survive by eating energy-rich food and creating warm clothes, and building shelters, using their surroundings. People living on very high mountains, such as the Sherpa people of Nepal, have adapted to live in their unusual environment. Their bodies use oxygen more efficiently to produce energy, which enables them to live in the foothills of Mount Everest, where the air is very thin and oxygen is less plentiful.

Technology can also help people survive harsh conditions. The Amundsen-Scott South Pole Station is a research facility at the most southerly point on Earth, where people from around the world live. Here, they study the history of Earth, the weather and the stars – all from a place with an average temperature of -49.5°C. In winter, the Sun never rises at the South Pole. Imagine that – being cloaked in complete darkness for 6 months of the year! Insulated buildings, electrical heating and ultraviolet lamps (to replicate sunlight) keep small scientific communities like these alive.

But there are some environments that humans can never survive in without help, for long. Underwater is one! It's incredible to think how dolphins thrive underwater. Unlike fish, who breathe the oxygen trapped as tiny bubbles in water, dolphins are mammals, and breathe air just like we do. They come regularly to the surface to grab a gulp of air through the blowholes (like nostrils) on the top of their heads! They can hold their breath for a lot longer than we can because of these blowholes, which can be sealed up when they are underwater. Our bodies haven't adapted to survive being in, or under, water for long periods of time.

But many extremophiles have. In fact, they love to live wherever nothing else can! Bacteria, in particular, are good at adapting to extreme environments.

Deep in the ocean, scientists have discovered bacteria living in boiling hot volcanic water. In Antarctica, they can survive in the frozen depths of lakes under almost a kilometre of ice. Some types of bacteria have developed special chemicals that stop their cells from freezing, so they can live at these extraordinarily low temperatures.

Other organisms have adapted to life-threatening environments by forming friendships with other living things. Have you ever seen a flat-looking green, grey or fuzzy, white thing growing on a rock? This is called a lichen. It forms when fungi grow together with algae or bacteria, a partnership that allows each species to flourish in otherwise impossible environments. Lichens have even been found on rocks in Antarctica, the coldest and driest place on our planet.

Tiny creatures can also survive in acid or alkaline (the opposite of acid) lakes, which would burn human skin. They even thrive in the Pacific Ocean's Mariana Trench, which is 11 kilometres deep at the lowest point. Here, the pressure is so high that the weight of the water would crush a car in 1 second. It's so deep that sunlight cannot reach the bottom. Amazingly, not only bacteria survive here, but also some species of fish and shrimp!

Isn't it astonishing how life takes hold in such mind-boggling environments? It makes me wonder where else in the universe life could emerge ...

The FUTURE of life on EARTH

Can you imagine Earth with flying dinosaurs and cars travelling in space? Well, actually, this describes the world we live in today!

Wait a minute, there are no cars in space, are there?

Yes! You may have heard of the rocket company SpaceX, which flies people and supplies to the International Space Station. In 2019, the company launched a car into orbit around the Sun as a publicity stunt. Along with its 'driver' – a mannequin wearing a space suit – it is now cruising through space, having completed nearly 2 orbits of the Sun. Where is the flying space car now? You can find out for yourself by checking out whereisroadster.com.

Alright, so there *is* a flying car in space. But there's no such thing as flying dinosaurs, is there?

You're not going to believe this, but fossils found in Germany and China show that small, feathery, winged dinosaurs existed around 150 million years ago. They evolved from large 2-legged dinosaurs called theropods, which included *Tyrannosaurus rex* and velociraptors. At first, these little dinos used their wings to scoot around quickly, or to glide to safety from a tree when faced with a predator.

Around 66 to 100 million years ago, these feathered raptors had evolved into a wide range of bird-like dinosaurs. Many of these (including those that lived in trees) were wiped out when their homes were destroyed by the asteroid impact we talked about before. Miraculously, some species (those living on the ground) survived – they are the distant ancestors of all birds that live on our planet today. Since then, more than 10,000 species of birds, from tiny wrens to mighty eagles, have emerged. Each of them can count T-rex as one of their distant relatives.

If Earth had such an exciting past, what can we expect to see in the future?

The continents have been moving since Earth's crust solidified. Our planet's outer layer is made up of many separate fragments, called tectonic plates, which float above the molten layers of rock below. As the plates shift, the land and oceans move, volcanoes spew, deep ocean trenches form, earthquakes rumble and colliding plates crumple up to make mighty mountains or islands.

The tectonic plates are still moving today, with Australia shifting north by up to 7 centimetres every year. In 100 million years' time, it's thought that Australia will have swept through Indonesia and South East Asia, and collided with southern China. Central America is likely to break away from South America, and Africa to slam into southern Europe, creating huge mountain ranges.

What will these changes mean for the plants and animals that populate our planet?

We have already seen how adaptable life is, and how it has varied throughout Earth's history. From microscopic single-celled creatures, to plants and dinosaurs, to life as it is today. Amazing things have evolved, like hearts, lungs, flowers, eyes and wings.

If dinosaurs developed the ability to fly, what remarkable skills could our planet's creatures have, millions of years in the future? Could we see a feathered bear with mind-reading powers? A wombat with x-ray vision? Might future humans even develop the ability to fly?

In 100 million years' time, human beings almost certainly won't exist – at least not in our present form. As a species, *Homo sapiens* (that's the scientific name for modern humans) have only been around for a quarter of a million years.

Over time, we have become smaller, less hairy, and our teeth and jaws have become less prominent. You might be surprised to learn that our brains are shrinking too! They are now the smallest they have been at any time during the past 100,000 years. Our skin, eyes and hair have also changed as we moved from our likely origins in Africa to every region of the world. As Earth's environment continues to change, it is likely that we will further evolve and adapt to our new conditions.

And we will need to do so quickly. In the past 200 years, humans have burned vast amounts of fossil fuels (that's oil, gas and coal). They are called 'fossil fuels' because they are made from the remains of plants and animals that have turned to sludgy carbon-rich mud. The mud is compressed over time to form coal, or heated by Earth's magma to form oil and gas. It's the burning of this fossil material that releases carbon dioxide into our atmosphere. This is rapidly warming the planet, causing polar ice caps to melt and sea-levels to rise. Global warming is causing more severe droughts, floods and bushfires.

Some places on Earth may become too hot for people to live in. Many coastal areas will become permanently flooded and unliveable too. Inland areas may become too dry to grow crops. We may need to combine our food and water resources as millions of people migrate across regional and national boundaries to escape these changes.

Humans might grow smaller muscles and longer limbs and develop new ways to keep cool. Can you imagine strange future people with larger ears, big hands or flaps of skin that help them to lose heat?

We are extraordinarily lucky to have this planet and to share it with such a colourful and exciting array of life forms. Could other planets in other galaxies have similar luck? I hope we live to find out!

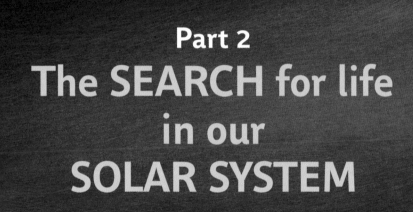

Part 2
The SEARCH for life
in our
SOLAR SYSTEM

LONGING for
lunar LIFE

If you look up into the night sky tonight, there is a good chance that you will see the Moon. It's big and bright, and the closest celestial object to Earth.

The Moon is Earth's only natural satellite. Together, the pair glide gracefully around the Sun once every year. The Moon swims around Earth every month, like a whale calf shadowing its mother.

The Moon may once have been a part of Earth, before an ancient planet called Theia is thought to have collided with our planet around 4.5 billion years ago. A huge cloud of rocks and dust was blasted into space, then pulled together by gravity. This gradually settled down, forming our Moon as we know it.

So, if the Moon was once a part of Earth, are the pair twins? Strangely enough, the Moon is a *very different* sort of place to Earth.

For starters, the Moon has virtually no atmosphere. Its gravity is far weaker than ours, so its lighter gases, like hydrogen and helium, quickly escape into space. Unlike Earth, the Moon also has a very weak magnetic field. That's because the Moon lacks a large melted iron core which, in Earth and other planets, creates a magnetic field as it sloshes around. This magnetic field is what protects our life-giving atmosphere from being slowly stripped away by the powerful rays that constantly zap our solar system.

Because the Moon has no atmosphere, there is no thermal 'blanket' to keep its surface warm. As a result, the Moon has extreme temperatures (from -170°C in the darkness to 120°C in the sunshine). It is barren and lifeless – so not like our Earth at all.

That's not to say we haven't searched for life on the Moon. People have been fascinated with this possibility for years.

In the 1800s, a German astronomer claimed to have spotted a city on the surface of the Moon. What he'd really seen was a series of cracks in the floor inside a large crater. He so badly wanted to believe that they were walls, built by an alien civilisation, that he started to concoct fantastical stories. The truth was discovered when astronomers with bigger telescopes and a more rigorous scientific method came along. One of the key rules of science is to let the evidence speak – just because you believe something, doesn't mean it's right without more testing.

Telescopes are great, but spacecraft are by far the best way to explore another world. In 1968, the first Earthlings used a spacecraft to get close to the Moon. Do you know who they were?

'Was it Neil Armstrong and Buzz Aldrin?' I hear you ask.

Actually, it was not.

The first beings from our planet to circle the Moon were 2 tortoises. The intrepid explorers flew aboard the Russian spacecraft Zond 5. They swung once around the far side of the Moon, before returning safely to Earth after a week-long mission. Although they didn't reach the Moon's surface, their visit was a giant leap towards the (slooooooow) exploration of the Moon and the search for life.

Meanwhile, the United States' National Aeronautics and Space Administration (NASA) was also shooting for the stars with the Apollo space program. Almost a year after our plucky tortoises left Earth to explore our solar system, 2 human astronauts aboard the spacecraft Apollo 11's Lunar Module managed to land on the Moon – Armstrong and Aldrin had quite literally been transported to another world.

But there was no swimming in the Sea of Tranquility (the place on the Moon where they landed) for Neil and Buzz. The Moon's 'seas' are dry, barren remains of volcanic lava flows from eruptions that happened billions of years ago. The only water found on the Moon is small amounts of ice mixed into the lunar soil at the poles.

Human exploration of the Moon was an important step in searching for life beyond Earth. More than 380 kilograms of lunar rocks were brought back to Earth from the Apollo missions. They were carefully studied by scientists for evidence of tiny living things like bacteria, fossils and basic organic chemicals (the building blocks of life). Sadly, they found no evidence of any of these things.

As a result, we are pretty confident that there is no alien life on the Moon and probably never has been. I think that's a pity. Imagine the fun we could have on holidays with our cosmic cousins!

Is there
LIFE on MARS?

One night, when I was 12 years old, my Dad and I went out into the garden after dinner to look at the stars. Among the blue and white twinkling lights, we noticed a strange looking orange 'star', small and round like a button. It was the planet Mars.

My mind was transported to another realm as I realised there was so much more to life than my home, my family and friends. In the night sky, I could explore a whole universe that is packed with mystery and wonder.

I began to imagine I was an astronaut sent to Mars, with a new world to explore. Would I venture up *Olympus Mons*, the mighty mountain almost 3 times higher than Mount Everest? Would I skip and bounce across the surface in my space suit, searching plains and caves and craters for Martian dogs or alien plants? Would I ever find life on Mars?

Astronomers have wondered about this question for hundreds of years. When strange dark patches were noticed on the face of Mars, some scientists thought they might be plants or forests on the surface of the planet. Others imagined they could be seas, lakes and swamps! It must have been so exciting to peer through a telescope and imagine Mars as a vibrant, watery and forested world.

Our understanding of the red planet improved when we started sending uncrewed spacecraft (with no people on board) to get a better view. It developed even further when we managed to land a spacecraft on the surface and study the soil, rocks and atmosphere of Mars. Hopefully, one day, people will fly there too.

The first successful visit by a spacecraft was NASA's Mariner 4, which flew past Mars in July 1965, taking the very first close-up pictures of the planet. From these images, scientists figured out that the dark patches were regions of rock, exposed by tremendous dust storms where the surface soil had been blown away.

In 1976, the NASA spacecrafts Viking 1 and Viking 2 landed in different places on the red planet. They used cameras and sensors to photograph the surface and study the atmosphere and soil of Mars. They found that the planet's air is very thin and dry, made up of 95% carbon dioxide, with only tiny traces of water. This discovery made scientists wonder whether life could survive on such a dry planet.

But hope quickly returned. Pictures from the spacecraft that have since orbited Mars reveal that the planet is home to mighty mountains, ancient dry rivers, dried out lake beds and even a grand canyon. These features made scientists ponder if Mars had liquid water in the past. Without a magnetic field to protect it, perhaps the water in Mars's atmosphere had been stripped away over billions of years by the radiation from the Sun?

We have also since learned more about what lies *beneath* the surface of Mars, thanks to the European Space Agency spacecraft Mars Express. Its radar detected a huge frozen lake, 100 metres thick, hiding under the surface of Mars! Scientists believe that the ice fell as snow millions of years ago, when the planet's axis was more tilted than it is today (giving less sunlight in winter) and the seasons were more extreme. Imagine making a snowman on Mars!

The discovery of underground ice changed everything. Perhaps there could be liquid water on Mars; maybe there are regions where life (whatever that might look like) could exist!

The excitement increased to fever pitch when possible channels of salty water were spotted flowing down the edge of a crater during a period of warm weather.

The questions keep building. Are there seasonal rivers or lakes yet to be identified on Mars? Could microbes (tiny life forms) have survived in the past, or still live in these regions today?

Mars rovers are being used to hunt for the answers to those questions. These self-driving electric vehicles have scouted several small areas of the red planet but, so far, have found no evidence of microbes (either living or dead). Future missions will look at new regions of the planet, specially chosen because of their potentially watery history. Maybe these searches will bring us the answers we are looking for?

The NASA Perseverance rover will investigate a dried-up lake and former creeks, deltas and shores, testing the soil for signs of ancient life. It will also trial a brand-new helicopter drone, called Ingenuity, which could enable future space missions to explore larger regions of the planet than ever before.

There has never been a more exciting mission to study the history of the red planet.

Is there life on Mars? Or was there life in the planet's past?

We may be about to find out!

SEARCHING the planets for LIFE

If the Moon is dry and lifeless, and Mars's ancient lakes and rivers are still being explored, what is the chance of life existing on a nearby planet? Let's strap on our space suits and take a ride through the solar system!

Mercury is the closest planet to our Sun. It is small (you could fit 18 Mercuries inside Earth) and has a very weak magnetic field. This has allowed radiation from the Sun to burn off most of its atmosphere.

Mercury also has veeeeerrrrrrrry looooooong days due to its slow spin rate and fast orbital motion. Almost 88 Earth days of continuous sunshine leaves daytime temperatures soaring to 430°C! At night, when the Sun finally sets, the temperature plummets to -180°C. Lucky we brought our space blankets on this trip.

For many years, scientists thought it was impossible for life to exist on a planet with such extreme temperatures. But looking closer, with recent missions like NASA's Messenger orbiting spacecraft, they found large regions of frozen water in craters near the poles of Mercury, which never receive direct sunlight.

With ice and gases (including hydrogen, helium, oxygen, sodium and potassium) bubbling out from beneath the surface of Mercury, scientists are now considering whether the right chemical mix for life could exist now or may have in the past. It seems unlikely, but could ancient fossilised microbes be hiding near Mercury's surface, just waiting to be found?

Since we have never landed on the surface, nor sampled Mercury's icy craters or soil, so much about this planet remains to be explored!

Okay, let's fire up our boosters and move on to Venus.

Remember the extremophiles? They would *have* to be extreme to survive on Venus. Although it is very similar to Earth in many ways, Venus's atmosphere has gone rogue with a runaway greenhouse effect, heating the planet to 471°C. That's hot enough to melt lead! Coupled with the intense atmospheric pressure (almost 100 times the pressure we experience on Earth), the chances of finding life on the surface of the planet are close to zero.

What about the atmosphere of Venus? This is thick, and filled with carbon dioxide, sulphuric acid and toxic gases like chlorine. Gas masks on, everyone!

There are some theories that extremophiles called thermo-acidophiles (thermo-acid-o-philes) – microbes that like heat and acid – could possibly live in the atmosphere of Venus.

These hardy little creatures are found on Earth in acidic lakes and where volcanic gases seep up through cracks in the crust. Some of these microbes can survive with only tiny amounts of oxygen.

Although we have no direct evidence of extremophiles existing in the atmosphere of Venus, the Japanese Akatsuki spacecraft currently orbiting Venus recently found strange dark features in the atmosphere that absorb ultraviolet radiation from the Sun. We have no idea what they are, but some have speculated that this could be large colonies of bacteria floating up high, like tiny birds soaring in the yellow Venusian sky. In 2020, scientists found a rare type of chemical called 'phosphine' in Venus's atmosphere. On Earth, this chemical is made by bacteria and deep sea worms, among other things.

Could bacteria survive up in the clouds? It's not such a wacky idea. In 2018, a team of Japanese scientists used an aircraft and scientific balloons to find *Deinococcus* bacteria that were floating 12 kilometres above Earth's surface. These bacteria are able to withstand large doses of ultraviolet radiation, which is handy in such a sunny spot!

So, the prospect of finding microbes in the atmosphere of Venus is definitely worth investigating. Engineers are drawing up plans for a space aircraft that could float through the atmosphere of Venus for up to a year, collecting data on windspeed, pressure, chemistry and even the presence of life above the clouds. Venus, we'll see you again soon.

Now, let's visit the 2 'gas giants', Jupiter and Saturn.

With no solid surface to stand on, there's certainly no animal- or plant-like life on Jupiter or Saturn. But what about flying microbes?

Probably not. The essential ingredients that make up all living things on Earth (sulphur, phosphorous, oxygen, nitrogen, carbon and hydrogen) are extremely rare on Jupiter and Saturn. Without these chemicals, it seems very unlikely that life *as we know it* could exist there.

Flying further out now, we reach the 'ice giants', Uranus and Neptune. These blue, gassy planets contain a more life-friendly chemical mix than Jupiter and Saturn. But the lack of an energy source like volcanoes, which spurred on life on Earth, leave us almost certain that the chemical reactions that create life could not take place there. What's more, the extreme temperatures and pressures on all 4 of these gas planets leave us without much hope that they could be home to carbon-based life similar to living creatures on Earth.

But we can't give up yet. There are still so many places to search. Are you with me?

Yeah!

Let's visit some of the icy moons of Jupiter and Saturn.

Icy MOONS

If millions of microbes are partying below the sea ice in Antarctica, is there any reason why we shouldn't find these little critters elsewhere in the solar system? Perhaps in other icy oceans, if they exist ...

Meet Ganymede, Europa, Callisto, Titan and Enceladus – some of the icy moons of Jupiter and Saturn. They are all named after Roman gods, but they have something else in common; scientists believe they may hide vast oceans of liquid water beneath their surface.

We can't see these oceans directly because, well, they are covered in ice. But scientists have found evidence that beneath the ice is a dark, sloshing world of salty water. They did this by measuring the effect the hidden oceans had on the magnetic field of the moons.

Ganymede, Jupiter's largest moon, probably isn't home to organic life even if it does have liquid water. Scientists think that the high pressure on its ocean floor may stop any hot-water vents bringing the nutrients needed to build life into the water.

Another of Jupiter's moons, Europa, is much better equipped to create and house alien microbes. Water is crucial to life as we know it, and Europa has a lot of it. Scientists think there may be twice as much water on Europa as there is in all the oceans on Earth! Although its surface is covered in ice, underneath, it is warmed by the friction of the tides caused by the gentle tug of gravity from its gigantic neighbour, Jupiter. These melt the ice and create what could be a perfect environment for life to thrive.

Europa's smaller sibling Callisto also has an ocean, around 250 kilometres below the surface. This ocean is probably cooler than Europa's, and less likely to support the chemical reactions needed to develop large organic compounds.

Enceladus (en-cellar-dus) is a moon of Saturn. Beneath its cracked, glassy, frozen surface lies a vast ocean that could provide a suitable home for living things. The Cassini spacecraft, a joint mission between NASA, the European Space Agency and the Italian Space Agency, spotted hot water, ammonia, salts and organic compounds from this ocean spraying out into space from hot vents, called geysers, in the ice. If life on Earth began in hot springs, what might that mean for Enceladus? This moon is a mouth-watering playground for alien hunters.

Titan is the big sister of Enceladus and is Saturn's largest moon. It has rivers of liquid methane and ethane flowing across its surface and is thought to have an enormous ocean beneath its surface. Whether this ocean contains water, we just don't know. But in recent experiments, scientists shone ultraviolet light (found in sunlight) onto a chemical mix similar to Titan's atmosphere and, without adding any water, grew the chemicals that make up DNA.

Sadly, we can't just get out our space submarines and start exploring. These moons are very far away. It takes at least 3.5 years to fly to Titan, and once you arrive, your spacecraft would be buffeted by the gravity and radiation fields from the nearby planets, Jupiter and Saturn. Keeping astronauts alive would be impossible with current technology and, even without people on board, landing a robotic spacecraft would be difficult.

That's why we are exploring these worlds one step at a time. Two space missions are currently planned to visit these icy moons.

In the next decade, a planned NASA spacecraft, the Europa Clipper, will fly low over Europa, Ganymede and Callisto, Jupiter's 3 largest moons, to take photographs and study the atmosphere. It will search for hotspots in their icy surfaces (like hydrothermal emissions or geysers) and uncover their watery underbellies using radar, magnetic and gravity measurements.

The European Space Agency is also planning a mission to the icy moons. JUICE (the JUpiter ICy moons Explorer) will cosy up to Jupiter and our favourite trio, Europa, Ganymede and Callisto, examining their oceans and atmospheres.

These missions will make exciting discoveries about what is hidden beneath the surface of some of the mysterious frozen moons of Jupiter and Saturn. Will they find evidence of organic chemicals venting up from warm pockets of ocean? And if they did, what would that mean?

We will only know *for sure* whether life has developed on these icy moons when we finally land a spacecraft on their surfaces and probe below the ice.

Will we find schools of fish or sharks swimming freely? Uncover Saturnian sea snakes or Jovian (Jupiter-dwelling) jellyfish? Or come face-to-face with some completely 'alien' creature we've never seen before? Will we discover aquatic forests that use energy from cosmic rays to grow, or underwater cities of intelligent ocean dwellers who only venture to the surface on rare expeditions?

Until we develop the advanced spacecraft capable of surveying distant worlds, landing on icy moons and drilling down into subsurface oceans, we can only dream of the mysteries lying beneath the icy moons of the outer solar system.

However long it takes, we will get there. Because human curiosity is a force too great to resist.

Are YOU an ALIEN?

We have so far assumed that life on Earth *began* on Earth, when chemicals mixed in hot ocean vents or on the shores of volcanic lakes.

But scientists have also considered another intriguing possibility.

Could life on Earth have come from space? In other words, are you and I actually *aliens*?

Imagine this: it's 4.5 billion years ago and the solar system is just forming. The planets have gathered material from space like giant snowballs picking up powdery flakes. A swarm of lonely leftover rocks are hurtling around the Sun like doomed messengers.

One icy asteroid, in particular, is on a collision course with our planet. Warmed by the Sun, a combination of water and organic compounds combine to assemble the first tiny living things on the asteroid. With an almighty impact, it crashes into Earth. Fragments scatter across a thousand kilometres, spilling precious living passengers across the ground and seeding life on our planet.

Another theory is even more exciting.

Hold onto your hats here, but we may just be *Martians*.

Incredibly, more than 250 meteorites from Mars have been found on Earth. These rocks are literally bits of Mars, thrown off by asteroid impacts, which have travelled more than 75 million kilometres across space to land in someone's backyard. How cool is that.

Some believe that meteorites like these may have brought ancient Martian life to our planet. Scientists have dissected these Martian meteorites, hoping to find their holy grail: an ancient fossilised microbe. The rocks studied so far have the organic chemical mix required for living things, but no certain signs of fully formed life have been found, yet.

If we do find microbes in a Martian meteorite – it's not impossible that these tiny 'Martians' took hold on our watery planet, thrived and evolved into the vast and varied living zoo we see today.

So if you've ever thought your brother, sister or friend might be from another planet, you could be right!

The idea that 'seeds' of life travelled across our solar system is intriguing, and it leads us to imagine an even more ancient possibility for the origin of life.

Could those little grains of life come from an even earlier time? From another star that existed before our Sun was even born?

In 2017, a strange and mysterious rock called 'Oumuamua (Oh MOO-uh MOO-uh) invaded our solar system. Its name in the Hawaiian language means 'first distant messenger'. Unlike a regular space rock (shaped like a potato) it's 400 metres long and thin like a slab of chocolate. To our eyes, the weird object was reddish in colour and had dark patches on its back. Scientists found it had no water and it didn't erupt in streams of gas like a comet would when it came close to the Sun.

This intriguing guest rocketed into our solar system and swung around the Sun, its orbital speed and trajectory proving that it was not from the outer solar system but instead it was an intrepid visitor from deep space! This was the first time that an interstellar voyager had been seen from Earth.

'Oumuamua's distant fly-by of Earth was only brief. It is now shooting off past the orbit of Neptune and leaving our solar system, returning to the depths of the galaxy.

What 'Oumuamua teaches us, more than anything, is that interstellar contact via rocky bodies is possible, at least over very long periods of time. And if tiny microbial lifeforms can hitch a ride on rocks like these and survive the depths of space, maybe there *could* be life somewhere else in the cosmos.

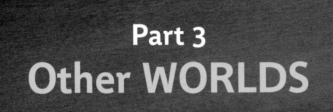

Part 3
Other WORLDS

Exoplanets

When I began my astronomy adventures as a 12 year old, astronomers had just a few weeks earlier discovered the *very first* planets around another star.

It was January 1992, and scientists were studying a weird star called PSR B1257+12 (let's call her Bernie for short). This small, but mighty, star was born out of a cosmic collision between 2 compact, cool stars called white dwarfs. She is 10 kilometres across (the size of a country town), spins at around 10,000 revolutions per minute and is a trillion times denser than lead. It's hard to imagine a more powerful little machine. If you crammed every human being alive into a thimble, that's how dense the star's material would be.

Bernie doesn't shine in every direction like a normal star. Instead, she releases energy in 2 torch-like beams, which flash more than 1000 times per second as the star spins around. Stars like this are called pulsars, because they have a regular pulse, a bit like a heartbeat.

That same year, in 1992, the scientists observing Bernie detected irregularities in her pulse and, like all good doctors, they wanted to investigate. It turned out that something invisible was tugging on Bernie's torch beams, causing them to wobble ever so slightly. What's more, the wobble had a very regular rhythm.

Scientists did the maths and figured out that this regular movement must be caused by a pair of planets orbiting the pulsar. These
hidden worlds were called Poltergeist and Phobetor. (These are their
real names, by the way, unlike Bernie, which is a nickname I made up!).

The scientists were also able to figure out what those planets
might be like. They are about 4 times heavier than Earth and quite a bit
closer to Bernie than Earth is to the Sun. Their proximity to the pulsar
means they would bake in sweltering temperatures. Any atmosphere
they once had would have been stripped away by energetic particles
streaming from Bernie the pulsar.

That wasn't the end of the discoveries. Another planet, called
Draugr (draw-ger) was found orbiting Bernie in 1994. This 'first family'
of extra-solar planets (or exoplanets, for short) marks an important
achievement in scientific history. After studying the stars for tens of
thousands of years, we humans have finally realised that our solar
system is not unique. It is not the only place in the universe where
life might exist! This opens up a new realm of possibilities, including
the chance of finding other planets like Earth out there. Could they
be teeming with life? Will we ever visit these planets, or speak to their
inhabitants? The questions are endless.

Thanks largely to the Kepler Space Telescope, which orbited the
Sun for more than 10 years and observed thousands of stars, we now
know of more than 4200 exoplanets. They come in many 'flavours',
including gas giants like Jupiter and Saturn, icy worlds like Uranus and
Neptune, rocky planets like Mercury, Venus, Earth and Mars and lava
worlds – hellish globes that are so close to their stars that their surfaces are bubbling cauldrons of melted rock.

And that's just the planets we have found in our corner of the
galaxy. With an even more powerful telescope, we might expect to
find planets throughout the universe. What might they be like? Could

some of them host life? Although we don't know the answers to these questions yet, we can make an informed guess.

According to data from the Kepler Space Telescope, as many as 1 in 5 Sun-like stars could have an Earth-like planet that enjoys a similar amount of sunlight as we do on Earth. Since there are an estimated 28 billion Sun-like stars in the Milky Way, that means we might expect to find as many as 5 billion rocky planets orbiting Sun-like stars in a region where liquid water could possibly exist on the planet, just in our home galaxy. Liquid water is required for life on Earth, so we guess it might be a necessary ingredient for extraterrestrial life, too.

Since there are probably about 2 trillion galaxies in total, there are likely to be around 10 billion trillion (10,000,000,000,000,000,000,000) Earth-like worlds in the universe. Let's meet some of them.

The HUNT for EARTH-LIKE worlds

Many scientists are interested in the hunt for habitable alien worlds. But what makes a planet suitable for life?

The truth is, we don't know. It's a big old universe out there, and we're not expecting to hear aliens knocking on the door tomorrow. That's why scientists have to make some educated guesses about what life might look like, and where it might live, so we can narrow down the search.

When looking for life in the universe, we usually start with places similar to Earth. We seek a rocky planet in the 'habitable zone' around its star, where liquid water could flow on the surface. That means the planet should be at just the right distance from its star, in what we call the 'Goldilocks Zone', where the surface of the planet is not too hot and not too cold (just like Goldilocks's porridge!).

We expect that living creatures, like us, need protection from dangerous extremes of temperature, atmospheric pressure and radiation. That means the planet can't be too close to its star, or it would be stretched and squashed by gravity and blasted with radiation. Yikes!

Over the past 15 years, several Earth-like planets have been discovered orbiting other stars. So, what are the most exciting, potentially alien-hosting worlds we have found so far?

One contender is Kepler-442b, a rocky planet orbiting a cool, orange star. It is in the Goldilocks Zone, so liquid water could possibly exist on its surface. But the planet is only 2.9 billion years old – that's 1.7 billion years younger than Earth. When Earth was that age, only basic single-celled and multi-celled organisms lived here. Our planet's animals, plants and fungi had not yet developed. That means even if life *has* begun on Kepler-442b it might be a world of microbes, not complex or intelligent creatures like you or me.

Another exciting Earth-like planet is Kepler-62f, the outermost of 5 (probably rocky) worlds in an alien solar system around 990 light-years from Earth. Kepler-62f is slightly larger than Earth, its year is shorter (at only 267 days) and it lies a little closer to its star than Earth does to the Sun. That's not to say the planet would be too hot for life though, because its orange star is fainter than the Sun.

Kepler-62f is almost twice as old as Earth. That's pretty exciting, because if the planet is habitable, and there are creatures living there, they will have had far longer to evolve than we have, here on Earth. Could complex, intelligent beings live on Kepler-62f? It is possible – and there's even a chance that they could have developed technology far more advanced than our own!

Imagine the scene: giant sea creatures bigger than a blue whale glide through the planet's waters; vast forests of alien ferns and trees growing wild, their oversized leaves collecting the precious sunlight from the dim star that Kepler-62f orbits, creating energy for the plants to live; flying mammals using their huge wing spans to overcome the extra 30% gravity on their oversized world. Complex, intelligent beings living among this beauty, exploring their solar system in fast and efficient spacecraft.

All this could be happening right now, in our neighbourhood of the galaxy. We can only speculate what mind-blowing worlds exist deeper in the cosmos.

I can't wait for a time when we can explore Earth-like worlds in person. But our current rockets are far too slow. The New Horizons spacecraft (our fastest interplanetary probe) would take around 20 million years to travel to Kepler-62f and Kepler-442b. Even if we could travel at the speed of light, it would take more than a thousand years to reach these planets.

Unfortunately, as far as we know, nothing can travel faster than the speed of light. When you are already moving really quickly, it gets harder to speed up. At the speed of light, you become infinitely heavy and can't move any faster. Weird or what?

This is why it's hard to move something heavy like a rocket even *close* to the speed of light, because it takes so much energy. If we *could* create a rocket that could travel at the speed of light, it would only take us 4.2 years to fly to the nearest exoplanet. (That's Proxima b, a small planet orbiting a red dwarf 'flare' star, called Proxima Centauri, that blasts it with bursts of radiation like an angry dragon).

With faster rocket technology, who knows what exotic neighbours we could meet? For now, we rely on telescopes to bring us closer to uncovering evidence of life in the cosmos. Our best option is to find Earth-like planets closer to home, so that we can fly there and back during a human lifetime.

Water BEARS
in SPACE

Some of Earth's creatures are extremely resilient and can survive in the dangerous environment of outer space.

In 2014, astronauts attached a box containing a variety of microbes, moss, fungi, lichens and algae onto the outside of the International Space Station. It was exposed to the harsh environment of space, with no access to oxygen.

These brave little critters experienced wild temperatures fluctuating between -157°C and 121°C as the spacecraft moved from bright sunlight into total darkness. They endured long-term exposure to ultraviolet radiation from the Sun and dangerous fast-moving particles called 'cosmic rays'.

The microbes were kept in outer space for about 18 months, before returning to Earth for study. Scientists were amazed to find that many had managed to survive.

Other experiments found that the hardiest of microbes, the *Deinococcus* bacteria, can persevere for many years in deep space, as long as they are protected from ultraviolet radiation. Those living inside rocks or in 'colonies' (or groups of their buddies) were happy as Larry. These microbes are tougher than we thought!

It gives scientists hope that living things might be able to endure the journey between planets. Whether that journey is taken on a meteorite from Mars to Earth, or from an interstellar rock like 'Oumuamua, the idea that life has travelled between planets remains possible.

What about other moons and planets? Could microbes exist there? We think they might!

Scientists have carried out several 'Martian' experiments in a special laboratory. Microbes were kept in soil and air identical to that on the red planet, to see whether they could survive. Most microbes died, but in damp soil that was rich in nutrients and protected from ultraviolet radiation, some managed to stay alive.

We haven't yet tested whether microbes can live on the real surface of Mars, but there are plans for humans to explore the red planet in the next 10 to 20 years, so we may soon have some answers. As well as testing whether bacteria from Earth can handle Mars, astronauts landing on the red planet will be hunting for Martian bacteria too. One of the prime targets for this search will be in the soil around the seasonal channels of what may be water seen running down the insides of Martian craters.

As we explore our solar system for signs of life, alive or dead, we must be very careful not to contaminate the surface of the worlds we are exploring.

Just as dirty hands can spread germs, spacecraft and rovers from Earth can spread bacteria and other microbial lifeforms to distant planets and moons. That's why modern space probes are carefully assembled and prepared for launch in 'clean rooms' (dust-free environments), to avoid sending Earth's microbes to other worlds. Space agencies were a little less careful in the past though, and we have probably already spread our germs to the Moon and Mars.

Nowadays, scientists go to great lengths to avoid leaving our proverbial mucky fingerprints all over other planets. The Cassini spacecraft, which studied Saturn and its moons from orbit, was deliberately steered into Saturn in 2017 when it reached the end of its mission. This was to avoid the spacecraft accidentally colliding with one of Saturn's icy moons. Imagine if we spread Earth bacteria across these pristine worlds – later, if we ever went back there, we'd never be able to prove whether any microbes that were discovered came from Earth or from Enceladus!

But things don't always go to plan. In 2019, an uncrewed Israeli spacecraft malfunctioned and crash-landed on the Moon, spilling its consignment of tardigrades (otherwise known as 'water bears') across the lunar dust.

These tiny creatures, less than 1 millimetre long, have 8 legs, cute little paws and look like something out of a science fiction book! They generally live in water, but they also love moist places on land like mossy ground. You might have some in your back garden.

If they dry out, water bears go into a sort of hibernation called *cryptobiosis* and their bodies shut down until they get wet again. It's like they have been placed in suspended animation, frozen in time and just waiting for a passing shower. Tardigrades can live for up to 30 years in this dried out state if the conditions are right.

On the Israeli spacecraft, a crew of water bears were placed into a sort of suspended animation, where they were dehydrated to travel to the Moon. The idea was to preserve these creatures as a record of life on Earth (not the best idea, it turned out).

Hopefully, one day, we will return to the Moon and pick up the water bears from the crash site. It will be incredible to sprinkle on a little water and see whether they come back to life. If so, this will be a fascinating insight into how long microbes from Earth might survive in the harsh environments of other moons and planets.

Red DWARF planets

About 7 out of every 10 stars in the Milky Way is a red dwarf. These stars don't shine hot and yellow like the Sun, nor brilliant white like the supergiant stars that dominate our night sky. Red dwarf stars are squat, dim and completely invisible from Earth, unless you have a very powerful telescope. They are also notoriously grumpy, throwing occasional temper tantrums in the form of powerful 'super flares' of radiation before returning to their previously gloomy state.

There are a lot of red dwarf stars. Probably at least 58 billion in the Milky Way alone, and several of these are close to Earth.

Do red dwarfs have planets?

Yes, they do.

The Kepler Space Telescope has found lots of planets around red dwarf stars, with some in the 'habitable zone' or 'Goldilocks Zone' of their puny stars.

The closest to home is Proxima b, a planet around the nearest star to Earth, Proxima Centauri. Even though the star is only 4.2 light-years away (we could theoretically fly there within a human lifetime if we built a futuristic rocket), it isn't visible to the naked eye. We know a fair bit about it though, thanks to some very large telescopes.

If you visited Proxima b and had a wander around, your legs would feel a little heavier than usual. That's because Proxima b's gravity is stronger than Earth's, so you would feel like you were walking around wearing a big, heavy backpack.

Proxima b orbits its star 20 times closer than we do to the Sun, taking only 11.2 days to complete 1 orbit. That means a whole year on Proxima b would last just 11.2 days.

Imagine that, a birthday cake every 11 days! I could definitely get used to that.

Could there be life on Proxima b? Sadly, we think it's unlikely. Proxima Centauri is another flare star. That means that it regularly experiences massive stellar storms, where the star violently ejects hot gas into space and becomes 50 times brighter than usual.

The ultraviolet radiation given off by these flares is about 100 times greater than the lethal dose for the sturdiest of Earth-bound microbes. If the planet had an Earth-like atmosphere, its ozone layer would quickly be destroyed by these regular storms.

Not all red dwarf stars have such violent outbursts though. There are some more serene red dwarfs, like TOI 700. This small, cool star is just over 100 light-years from Earth and it has an Earth-like rocky planet called TOI 700 d. The planet sits right in the zone where liquid water, and therefore life, could exist!

Although it is safe from star flares, there is something odd about the planet. It always has the same side pointing towards its star. This leaves one side in the permanent glare of sunshine and the other in a state of constant, relentless darkness.

Could there be life on this peculiar planet?

Perhaps. We know from studying Earth's deepest oceans that living things can survive without light. So long as the temperatures are not too extreme (which would be helped if the planet has an atmosphere), maybe life could exist on TOI 700 d close to the boundary between day and night. Creatures there would enjoy a permanent sunrise and bask in the reddish-pink glow of their star.

I would love to visit and learn more about this potentially habitable world. Unfortunately, at 100 light-years away, it would take us many thousands of years to reach it in our current spacecraft.

So, we will not be visiting the planet any time soon. Our best hope of learning more about life on these very distant worlds is that we send super-fast spacecraft towards the stars with the knowledge that the next generations of scientists will learn from the results.

That's a pretty awesome thought!

HOLIDAYING on an alien WORLD

Imagine we are aliens, living on another world.

There are 6 planets in our solar system and at least 3 of them could possibly support life. We live on Zorg, a rocky planet with an atmosphere of nitrogen and carbon dioxide, with a little oxygen mixed in. Our 4-legged bodies use mostly carbon dioxide to create energy.

Our planet orbits 2 orange stars, which warm its surface to an average temperature of 60 °C. This is perfect for us. It doesn't feel too hot, because our bodies are adapted to survive here.

Many other plants and animals live on Zorg, including several species of intelligent birds. On the land, vast sandy deserts stretch into the distance. In the cooler regions, towards the poles, are forests of kapko, a tall spongy moss. Its deep blue colour is quite beautiful.

Our neighbouring planets, Ela and Marna, are also inhabited, but by different creatures that evolved independently. Life began on these planets more than 7 billion years ago, which is why there is such a vibrant range of complex and intelligent creatures living there.

Ela is a world filled with dazzling, flowering plants, and giant, flying insects similar to butterflies. There are also large numbers of intelligent bee-like creatures called *apians,* who live in small families. They build their own shelters, use a complex form of body language and have formed societies where they help each other by producing the various foods and raw materials they need to live.

Marna is a cool, watery world, with intelligent ocean-dwelling creatures that live in homes constructed underwater close to the shores of seas and lakes, to protect them from predators. They speak complex languages similar to those of whales and dolphins, yet their lives are quite different as they have learned to farm fish, and no longer need to hunt.

These worlds used to seem very alien but, ever since the inter-planetary express was built, it has been possible to go on holidays to Ela and Marna.

When our fellow Zorgians first started to visit these fascinating worlds, they used to wear space suits to protect them from the uncom-fortable temperature and low atmospheric pressure there. Since the Zorgospheres (special capsules, like undercover holiday resorts) were built, we only need special clothing if we venture outside. And that isn't recommended on Ela, since the apians can get quite angry if they see you flying towards their precious flowers!

Trips to Ela are amazing. Being closer to the twin suns, it's a trop-ical getaway with very long days and short twilit nights. Luckily, we're used to that on Zorg, since having 2 suns means only a few hours of darkness every night.

Sightseeing on Ela means zooming around on small, flying bikes, taking advantage of the weaker gravity. Even through a thin pressure suit and helmet, the scent of the gigantic caroo flowers excites your senses. The massive 4-winged butterflies were scary at first, but they are harmless, and their colours have to be seen to be believed!

Holidays on Marna are more of an adventure. For starters, it's freezing cold! You have to wrap up in thick clothes above your pres-sure suit. Then there's the gravity, 40% stronger than on Zorg. As you land on the planet, you feel strangely heavy in your seat. Just getting up and walking down the steps from the spacecraft is quite difficult. You suddenly feel like a very sluggish creature. Luckily, there is very little need to walk for long distances on land.

Most of Marna is made up of ocean, and that's where the excitement lies. The highlight of most holidays on Marna is a thrilling cruise beneath the sea. The submarine vessels, specially designed with the comfort of Zorgians in mind, are pressurised to 1 Zorg atmosphere and calibrated with exactly the right mix of gases to enable us to breathe easily.

See-through viewing platforms enable us to watch the everyday goings-on of the magnificent Marna creatures as they work and play. We glide slowly past their homes and farms, watching as they get on with their secretive aquatic lives. This must be what it feels like to be a fish!

The submarines are designed to absorb the radar signals these creatures use to 'see' underwater, so the locals don't even notice we are there. It's best not to frighten them, to keep this a safe and sustainable holiday destination.

Is this a realistic picture of an alien solar system? It just might be.

Back in the real world, the TRAPPIST-1 exoplanet system is only about 40 light-years away. It has 7 planets orbiting a faint red dwarf star. Three of the planets are in the habitable zone! A future space mission (travelling close to light speed) could explore all these worlds, beaming results back to Earth in less than 100 years.

In the meantime, astronomers continue to search for radio signals from the planets, in case we intercept interplanetary communications. None have been found so far.

We'll be keeping our eyes on TRAPPIST-1 and its 7 planets. Imagine the holidays we could have there!

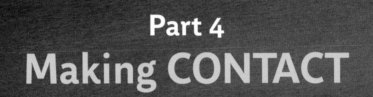

Part 4
Making CONTACT

Flying saucers,
SPRITES
and UFO battles

Do you find the thought of aliens interesting, or spooky? For me, it depends on how I think about what they might be like. If we're talking about microbes on asteroids, or graceful creatures swimming deep in the oceans of Europa, I'm curious, and want to know more. But if it's flying saucers hovering above dank forests and abducting people, I'm out!

Is there any truth in the reports of sightings of extraterrestrials visiting us in flying saucers?

Tales of strange, glowing balls of light, or fire in the sky have been around for a very long time. In 1561, in the German city of Nuremberg, a local newspaper carried a story of giant luminous arcs, crosses of blood-red light and other shapes dancing across the sky. Some modern commentators have dubbed this a 'UFO battle' in the skies.

Reports of UFOs became common when humans invented aeroplanes. Sightings of balls of light called 'foo fighters' were common in World War Two. People who saw them said that these objects could move direction quickly and often seemed to follow aircraft.

A cascade of UFO sightings and claims of flying saucers followed, mostly in the United States. In 1947, a pilot called Kenneth Arnold (and several witnesses on the ground) claimed to have seen a line of thin, circular flying objects over the north-west of the United States. There was considerable interest in the media, and many other sightings of strange objects were reported by others. The wreckage of a supposed 'flying saucer' was found in July that same year by locals near the small town of Roswell, New Mexico.

Can science explain these incidents as more mundane events? Or have intelligent creatures from other worlds really visited our planet?

Fiery flashes and incandescent balls of light can be caused by thunderstorms. They include blue or purple plasma discharges, often seen from aircraft, called 'St Elmo's Fire', red flashes seen above thunderstorms called 'sprites' and mysterious floating balls of light formed during electric storms, called 'ball lightning'. No aliens here.

Can we explain the 'UFO battle' in Germany? Probably. It was almost certainly a 'sun dog'. These beautiful light shows are created when light is reflected into glorious rainbow circles and arcs by icy, high-altitude clouds.

And the debris found at Roswell? Not aliens, sadly. It was just a US Air Force high-altitude balloon, which crash landed in the area.

What about the flying saucers seen by witnesses on the ground and from aircraft though? Could they be life from outer space?

That is something we can't explain.

A large number of pilots have seen fast-moving dots of light, either directly with their eyes or popping up on their radars. Nobody can clarify what caused each and every one of these events. They remain a mystery.

But aliens? It seems like quite a coincidence that most UFO and flying saucer sightings suddenly happened during World War Two, a conflict waged mainly by air. All sorts of experimental aircraft and rockets were being tested from the 1940s onwards, particularly in Europe and the United States. These were top-secret military projects, and nobody was allowed to talk about them. It's possible that these 'UFO' sightings were simply secret aircraft, drones and rockets. But we don't know for sure.

What about astronauts travelling in space? Have they ever seen an alien spacecraft?

Astronauts have reported seeing objects floating in space outside their capsules. The internet is full of conspiracy theories about aliens flying around Earth, but the truth is more mundane. These were found to be just pieces of 'space junk'; discarded boosters and equipment jettisoned from our own space activities. So, litter, not living beings.

People love to believe that every unexplained light or shape they see in the sky might be intelligent beings from another planet. It's an exciting thought that one day we might be able to meet or speak to creatures from another world. But science is based on *evidence*, which means having proof that something is true. With UFOs, we have no proof of what they are, and certainly no evidence that they come from other planets. It's always best not to jump to conclusions, even if that means saying *we don't know.*

Nowadays, almost everyone carries a smartphone with a high-quality video camera. If UFOs really *were* carrying aliens from another planet, we should one day have good, high-quality footage of them. I won't hold my breath!

Is ANYONE listening?

Since the invention of radio transmitters, humans have been leaking radio noise into space. Our airport and weather radars, our mobile phone and TV transmitters, are constantly blasting information about our civilisation in all directions. If there are smart aliens living on Proxima b, our neighbouring exoplanet, for example, they could easily detect these signals using a radio telescope.

Our communication with alien species may already have begun without us realising it! Scientists on Proxima b might already be monitoring our broadcasts, watching our TV shows and learning about our languages and customs. If they have seen our nightly news programs, featuring conflicts and wars, they might think *we* are very scary aliens. Maybe they're happy just listening, and don't want us to know they exist.

As well as these accidental messages to any extraterrestrial neighbours who might be eavesdropping, humans have also sent deliberate messages into space.

In 1974, the giant radio telescope at Arecibo, in Puerto Rico, beamed a bright radio signal towards the Messier 13, or M13, star cluster. The message contained information about the solar system, humans, DNA and our position in the Milky Way Galaxy. If any of the

100,000 stars in the cluster is home to intelligent beings who use radio technology, we might expect a reply in around 50,000 years' time!

A more recent broadcast targeting a nearby star is far more likely to get a response. In 2008, the radio dish in Yevpatoria, Ukraine, beamed more than 500 messages from Earth toward the star Gliese 581, which is only 20 light-years away. Its planetary system contains 2 potentially habitable planets, including Gliese 581c, which may be very similar to Earth. The signal is expected to arrive early in 2029, and if there is an intelligent civilisation that decides to reply, we could receive the message as early as 2049. How old will you be then?

The big question is: what would we do if we received a clear signal from extraterrestrial beings tomorrow, telling us where they live and what they look like?

There is a rulebook for the *dos* and *don'ts* of alien contact. Scientists are guided by an international agreement made in 1989, to say what should happen if an alien signal is discovered.

First, any signal that we 'think' might be aliens (and we find a lot of those) should be checked by several other scientists to make sure it's really another life form. We don't want to spread panic, just to find out it was a false alarm.

If a signal is confirmed to be from intelligent extraterrestrials then national governments, the United Nations and international scientific organisations must be informed.

The group of scientists that discovered the signal can then announce the discovery to the public. All information should be shared, so that everyone has a chance to see exactly what we're dealing with. Most importantly, *nobody should reply* to the message before governments across the world have had time to discuss exactly what we should do.

When 2049 comes around, if a message arrives from the folks from Gliese 581c, what information would you share with the aliens?

Would you send e-books, art, music or TV shows? Would you share information about science, or ask questions about the aliens' knowledge of technology and the universe?

Should we arrange to meet up? Or should we stay quiet, in case they want to harm us?

One day, we might have to make those decisions.

Playing FOOTBALL
with
EXTRATERRESTRIALS

Would aliens from other worlds be able to visit Earth? Could they land their, er, flying saucers or whatever, and saunter down a ramp to walk among us? Could we have a game of football with them?

It depends. Aliens' bodies are probably quite different to ours. The gravity, the atmosphere and the strength of the sunlight where they come from may be completely unlike those on our planet. If they were to visit Earth, they might feel heavy and sluggish, or bounce around like humans do on the Moon because our gravity might be weaker than theirs. They might even need to wear thick and heavy space suits to protect them from our atmosphere.

See – they seem less scary already!

The chances of slimy, green aliens with big eyes showing up are pretty slim. But what if we received a message from intelligent beings on a distant planet? If they gave us their address, could we go and visit – maybe, kick a ball about?

It is possible. Out of the 4200 exoplanets we know of, about 60 are close enough for us to visit in our lifetime. This means it might be feasible, one day, to leave Earth, fly across space at the speed of light or close to it, play football with the aliens, have a cold drink and a chat, then fly back to Earth.

The alien football part sounds fun, but the travelling not so much. Imagine spending 20 years in a small spacecraft, with no fresh air and only a few people to talk to – and that's just one way! Even if we wanted to spend our entire life cooped up in a tin can with no sunlight, drinking recycled urine (yes, really), we still don't have the technology today that would make it possible.

We would need to find an energy source to power a flight that could accelerate to 300,000 kilometres per second and then slow down again so that we could explore the planet. We would need engines and parachutes to land safely on the surface (not easy) and to blast off and fly home again.

Spaceflight would affect our health too. Astronauts need to be protected from dangerous cosmic rays. Weightlessness causes havoc with muscles, eyes and bones. A lack of natural light and air can be harmful. Space junk is a constant hazard, and the boredom would be terrible too. Imagine how big your jigsaw would need to be to last for a 40-year journey!

Excitingly, there are a few ways a journey to visit aliens might work.

If there is life on our solar system's icy moons, space tourists could be wriggling into their space suits and jetting off to the gas giant planets within the next 50 years. Even if those alien creatures just resemble microbes, or even water bears, who wouldn't be excited to climb on board a submarine and visit our nearest neighbours?!

If there's no other life in our solar system, maybe we'll be lucky enough to meet up with an alien civilisation among our closest stars. If they also had spacecraft, meeting halfway would reduce travel times. We could play our football match (yep, it's my favourite sport!) on a nearby planet or even in a specially designed space station with an atmosphere that suited both species. It seems unlikely now but, 60 years ago, so did the idea of humans flying to the Moon.

Our final hope of making friends with aliens is that *maybe we can travel great distances across the universe.* Rather than breaking the speed of light, we might discover secret tunnels through space and time. These are what we call 'wormholes', proposed by scientists as a way for humans to travel long distances through our cosmos. Although there is no evidence (yet) that they actually exist, we're still looking.

Imagine flying through a wormhole to a distant land, 2 billion years in the future. What incredible creatures and unfathomable technologies might exist then?

For now, this is just science fiction. Luckily, your imagination has no bounds.

VIDEO chat
with ALIENS

With our land-loving bodies and short lifetimes, visiting our alien neighbours may never work out in practice.

So why don't we just video chat?

The idea of sending and receiving messages from alien worlds is nothing new.

More than 100 years ago, scientists discovered that it was possible to transmit messages across long distances by putting an electrical current through a metal wire. This created invisible 'radio waves', which could be intercepted by another wire hundreds of kilometres away. This technology was revolutionary because it allowed messages to be sent instantly across the oceans. And it's still how we send mobile phone, television and radio signals today.

If we use radio communications here on Earth, maybe intelligent Martians are also transmitting radio signals to each other? Pioneering inventors experimenting with this technology in the early 1900s noticed unexplained 'static' (crackling noises) in their radio equipment and wondered whether it came from intelligent Martians. There was no evidence either way, but scientists in those days got very excited about the thought of chatting with alien creatures. Disappointingly, the static turned out to be from Earth, and not from Martians after all.

As radio transmitters and receivers got bigger and more fancy, scientists searched once again for Martian communications. In 1924, the United States declared a 'radio silence day' and all Earth-based radio broadcasts were turned off for 5 minutes every hour. *Shhhh! We're listening out for aliens!* They pointed huge radio dishes towards Mars, which was at its closest point to Earth for more than 80 years, to eavesdrop on Martians. Sadly, they didn't hear any messages from the red planet.

Modern alien hunting uses radio telescopes, which are usually giant metal dishes that can be pointed in any direction. We call these projects the Search for Extraterrestrial Intelligence (SETI). I used to work with the SETI team using the Parkes Radio Telescope in New South Wales. It was great fun getting the telescope ship-shape for alien hunters to use!

Telescopes around the world are now scanning the 1 million nearest stars to Earth, looking for radio signals that might be 'leaking' from an exoplanet. They are also hoping to find signals beamed directly into space – for example, communications with spacecraft fleets, or 'hello' signals broadcast by friendly aliens hoping to receive an answer.

Have we found anything?

During a routine night of observing at the Ohio State University radio observatory in 1977, a sharp spike of radio emission was detected in the direction of Sagittarius. The signal was 30 times stronger than the usual background noise and was recorded for only 72 seconds before the telescope moved on to another target.

The strange peak was only noticed later when Jerry Ehman, a volunteer at the observatory, checked the computer printouts from the previous night. When he saw how strong the signal detected was, he circled it and scrawled 'Wow!' on the printout. After that it was named 'The Wow! signal'.

Scientists around the world scrambled to take another look,

hoping the signal would still be there, or that it might come back. Despite scouring the area for years, it was never seen again. To this day, we still don't know what caused the Wow! signal. Was it a beacon from an extraterrestrial civilisation? Was it an 'air traffic control' tower for a Sagittarian Space Fleet? All we can do is keep looking.

On the 35th anniversary of the Wow! signal, the most powerful radio transmitter in the world, the Arecibo radio telescope, beamed messages and videos from Earth toward Sagittarius, just in case it was alien creatures who sent the original signal. Trouble is, it's unlikely that anyone will receive our message for hundreds, if not thousands, of years.

Radio signals, like light, are bound by the speed of light. For a phone call or video chat with a planet 1000 light-years away, you'd have to wait 2000 years to receive a reply.

If we do want to Skype our new alien friends, they'd better be somewhere close by.

Alien AMBASSADORS

The Sun is around 4.6 billion years old and has at least another 5 billion years to live. Let's imagine a future time, when life on planet Earth is far more advanced. A time when travel between stars is possible, and when creatures from different planets live beside one another.

It's 200 million years in the future. Human beings became extinct long ago. An intelligent species, let's call them Payons, now inhabit our planet. The Payons have harnessed renewable energy from wind, water and the Sun. They have learned to live in balance with their natural environment and have expanded their activities into space. This has enabled them to develop into superintelligent beings.

Long ago, the Payons put on their space suits and explored the solar system. They set up scientific stations in orbit around every planet and moon. They explored craters, mountains, volcanoes and subsurface oceans, learning a great deal about how the solar system began. But they quickly realised there was no life on these worlds, and nobody in our solar system to talk to.

Not to be beaten, the Payons sent a fleet of robotic spacecraft to nearby exoplanets. They used artificial intelligence to steer the spacecraft and explore these strange worlds. They extracted small amounts of minerals and metals from asteroids and moons along the way, making new fuel and making sure their spacecraft remained light enough to manoeuvre at high speeds.

Payon space probes focused on planets that were more than 5 billion years old. They figured this would increase their chances of encountering complex intelligent life, since the creatures would have had longer to evolve. They found that most rocky planets were life-less – they were simply too hot or cold, or their atmospheres were too thick, or thin.

Finally, after a few hundred years they stumbed upon a rocky planet teeming with creatures, just like Earth. And after thousands of years of searching, they came across an intelligent civilisation, known as the Brakk people, who like the Payons had learned to transport themselves into space and adapt their bodies to new environments. This space-faring civilisation was successful, because they lived on a very small, rocky planet where gravity was weak, and it was easy to launch into space.

Communication with the Brakk was difficult at first. The Payons had trained their intelligent spaceship computers to develop relation-ships with alien creatures. They showed the Brakk people realistic holograms of Payon society and technologies, translated into infra-red images since the Brakk's eyes could barely detect light.

Many Brakk people were frightened at first, and mistrusted the Payon visitors. If the Payons had travelled to Brakk and landed unan-nounced, they may have faced hostility, or even violence. But sending a robotic spacecraft first allowed the Payons to say hello and comm-unicate their knowledge and personalities via computer, without the Payons being there in person. This helped the 2 species form a strong relationship before a meeting was agreed.

Once communication was established, a team of Payon diplomats began the long journey to the distant planet, Brakk. In preparation, they genetically engineered their own DNA to adapt to the atmosphere and levels of cosmic radiation at their destination. That way, when they arrived on the planet, they could live almost normally. Diplomatic processes went well.

Soon, parties of Brakk began visiting Earth. They were treated like leaders, or ambassadors, and given a warm welcome and tours of our planet. Knowledge was shared freely, enabling both species to develop new technologies to make living and travelling in space safer, more comfortable and sustainable for everyone.

Nowadays, many Brakk and Payons live together on both planets. And more alien civilisations have been discovered, with advance parties travelling across the 4th quadrant of the Milky Way to meet and trade ideas. Peace is essential in all these interplanetary interactions. But creatures intelligent enough to develop sustainable practices are too advanced to fight and quarrel about resources like minerals and metals anyway. They all have everything they need from sharing the almost infinite resources of the universe. Now *that* sounds like a great way to live.

Part 5
What are aliens like?

Do ALIENS have big, GOOGLY EYES?

Life on Earth is astonishing. From parrots to penguins, panthers to platypuses, the animals that share our world are stunning in their variety. Whether it's bacteria living in a rock buried deep beneath the ocean, an extremophile bathing in a hot volcanic spring, or a lichen surviving on the windswept, icy plains of Antarctica, every organism has evolved an individual way of surviving on our planet.

Many of Earth's creatures look alien to us, like the weird Mexican amphibian, the axolotl, or the slimy green ribbon worm of south-east Asia. Look them up! As odd as they look to us, both are perfectly suited to their environment. The sheer number and variety of living things on Earth tells us that life on another planet would most likely be in very different forms to what we are used to and just as diverse as well.

Creatures living on comets, asteroids and meteorites would probably be tiny, microbe-like beings. With nowhere to hide, any larger creatures would be zapped by a constant rain of dangerous particles from space. These microbes may have developed special biological 'radiation shields' to protect them from cosmic rays, like in-built magic umbrellas!

Life forms on massive worlds may have extra strong legs, with thick bones to support them against the strong gravitational pull of their planet. Or maybe they would have no bones at all, like worms or slugs. Imagine that!

Aliens could be bigger than the mightiest mammoths, or tinier than the most miniature microbes. They might soar through the skies like eagles, or be anchored in the ground like trees. Imagine a weird world of intelligent plant-like aliens making tools by sucking up nutrients and '3D printing' the materials that they need through their cleverly evolved limbs.

Perhaps aliens don't have bodies made from carbon like we do but, instead, use silicon and other chemicals to fire their biological engines. In that case, living creatures might thrive in places we don't expect, like the methane rivers, lakes and seas of Titan, or on other icy moons throughout the universe. Maybe they don't have a solid body at all. Perhaps some living creatures might take a liquid, or gaseous, form.

Movies about aliens show them with big, googly eyes. Is that how they'd look in real life?

Eyes are very common in creatures on Earth. But on a planet with very little light, or on a comet or asteroid flying through deep space, there may be no need for living creatures to develop eyes at all. Aliens from dark worlds might rely on sounds, smells, vibrations or even infra-red radiation to navigate and communicate.

How would aliens eat? We humans simply open our mouths, pop some delicious sushi (or whatever it is you prefer) and digest the food in our stomach and intestines. Plants eat by sucking nutrients from the ground through their roots, and many gain energy from the air and sunlight via their own special solar panels – their leaves. This process is called photosynthesis.

Fungi eat plant or animal matter by injecting digestive enzymes into it and breaking it down outside their bodies, before sucking up the goodness. Yum!

Microbes eat in strange and varied ways. Some bacteria gobble energy from the Sun and air by photosynthesis, others use sunlight to absorb nutrients that they suck through their cell membrane (that's like the bacteria's skin). Other microbes called phagocytes (fagg-oh-sites) are hunters, wrapping themselves completely around other microbes and digesting their prey.

Could aliens get energy from starlight, or by absorbing food through their skin? There's no reason why not. Could they digest their food outside their bodies and suck it up through the air or through roots? Weird, but entirely possible!

Maybe they would get their energy in strange ways, not seen on our planet? Sure, why not! Aliens might get their strength from renewable sources like wind, water, nuclear power or volcanic eruptions.

Anything could be possible in the theatre of life.

Are aliens
INTELLIGENT?

We humans think of ourselves as the most intelligent creatures on Earth. That's because we're able to think in complicated ways about the world around us. We recognise patterns, learn, plan and solve problems. We even play Scrabble!

Humans have developed mathematics and science to understand more about the world. We create complex things like music, art, dance, stories and games to feed our souls. We use technology to communicate across the world, to make manual work easier, to fight diseases and to explore our planet and outer space. We even have machines to do the dishes! That all sounds pretty smart.

But humans are still learning. In our race to make life easier, we rarely consider what makes our lives truly *better*. And, as a result of our technological 'intelligence', we are now scrambling to save our natural environment from pollution.

Are aliens harnessing the power of technology without harming themselves and their environment? That would be a sign of advanced intelligence – and it might ensure that they lived long enough for us to find them, or for them to find us.

Have you ever wondered what a superintelligent alien species would be like? I have, and I want to know if they would like music. Any species that has developed advanced intelligence would need to have a complex way of communicating. Music could be a strong part of that. On Earth, birds sing and chirp, frogs croak and we sing and dance.

But what if aliens live in a place with little or no atmosphere? There might be complete silence, because sound waves need a fluid – that is, a substance that flows, like air or water – to travel through.

That's okay though, as there are many ways to make music. Aliens could communicate using radar, by movement or dance (waggling like bees) or by thumping the ground and absorbing vibrations through their feet. Imagine the spectacle of an alien choir, banging the ground in chorus!

It is worth considering that the most advanced alien species might not be using technologies that we can detect at all. Yet! As humans have progressed in our scientific knowledge, we have discovered previously unknown forces like gravity, magnetism and electricity, and invisible carriers of energy such as x-rays, radio waves, infra-red and ultraviolet radiation.

Given that there is so much more to learn about the universe, there are undoubtedly many more of these hidden scientific gems to uncover. If alien civilisations have developed for thousands or even millions of years longer than ours, it is likely that they would have access to forces of nature that we can't even imagine. To us, their abilities might seem like superpowers!

But aren't aliens curious, like us? Well, some people believe that curiosity is a sign of intelligence. But you might have heard of the old saying 'curiosity killed the cat'.

Although some groups of people through history have chosen a life of exploration and expansion, this behaviour often leads to struggle and conflict. Many human societies value their health, connection to culture and land, clean water and living in peace and stability above all else. If aliens are truly intelligent, and their culture has survived long enough for us to stumble upon their planet, then perhaps they are not exploring their solar system or beaming radiation into space, but instead are living quiet, peaceful and happy lives, not needing or wanting to be found by us at all.

Could ALIENS LIVE forever?

Do you want to live forever?

In the year 1800, the average human being lived to only around 30 years of age. That might sound old, but ask the grown-ups in your life how old they are and you might get a shock!

The average life expectancy has more than doubled since then. Today, on average, it is almost 73 years. But it is not equal across the world. In the Central African Republic, the average person lives to just 53. In Japan, the average lifespan is 84 years. The oldest person who ever lived was a French woman called Jeanne Calment who lived to 122 years! Now that's old.

Some of Earth's living things can live *much* longer than humans. Quite a few types of fish, including sharks, can live for more than 200 years. Scientists have found living trees that are more than 5000 years old – that's older than the Egyptian pyramids! Can you imagine what those trees must have experienced in their lifetimes?

One animal surpasses all other life on Earth in its ability to endure. In 1986, scientists found the skeleton of a sponge (an animal that lives on the seabed) that had lived for more than 11,000 years. Amazing!

They figured out its age in 2012 by studying the different types of oxygen (some with heavier atoms that others) in the sponge's skeleton.

From this, they estimated the temperature of the ocean when the skeleton formed, and since scientists know how warm the oceans were at different times in history, they could estimate the age of the sponge.

From the everyday to the complex, stupendous health inventions including soap, vaccines and medicines help us humans to stay healthier for longer. Since we're living longer than ever, could we one day live for 11,000 years, just like that sponge?

In recent years we have learned to adapt our bodies using technology, including artificial hearts and pacemakers and bionic arms and legs. Imagine what else we might develop in the future!

But it would take tremendous advances in science to make people live forever by simply replacing our organs with artificial ones. So many different parts of our body would need replacing. We would need a constant stream of operations and procedures to fit our new parts. An interesting idea, but it's not without its problems. It raises all kinds of philosophical and ethical questions, too.

Instead of using mechanical body parts to live longer, we *could* try to change our DNA (the in-built instruction manual in our cells) to slow down the ageing process. This is called 'genetic engineering'. Scientists have already done this in mice that had a disease that made them age faster than usual. They made changes to the mice's DNA that made them healthier and live 25% longer.

If we developed similar technologies and applied them to people, it is possible that we could live longer lives.

But should we be trying to live longer?

If people lived to 200 years of age there would be so many people, we might not have enough space, food or clean water to go around. That could cause more wars and conflict, and pollution and climate change could get even worse. So maybe we don't want to become everlasting cyborgs, but simply live a life in balance with our planet and our environment – and get better at sharing what we have.

Assuming a highly intelligent alien species has learned to share its resources and use clean, renewable forms of energy, could they develop sustainable ways to live forever?

Perhaps they could, particularly if they don't live inside fragile bodies, like we do.

Genetic engineering is one possible way for aliens to adapt to changing environments on different worlds, and possibly to survive long periods of travel between them. Although we don't yet have the technology to adapt our bodies to rapid changes, it is not impossible that a far more advanced species could have found a way to edit their DNA to morph their bodies in different ways.

A highly advanced species might choose to live in digital form. That means their brains – all their thoughts, personalities and emotions – could be saved on a computer. This information could be transmitted (at the speed of light) to different planets, moons or star systems. The aliens could even choose to upload themselves to artificial bodies, like robots, at different locations. Imagine living in digital form and transmitting your brain to distant worlds. You could go anywhere and experience anything!

Aliens could go one step further and upload all their personalities into a computer to create a single 'artificial intelligence' version of their species. In this way, the computer version of the species could leave their world, travel across the cosmos and explore new planets. It could survive the destruction of their planet by an asteroid, outlive any single individual and travel across the universe. All it would need is an energy source to sustain the spacecraft and its computer. If it broke down though ...

Do you like the idea of meeting an alien intelligence that lives as a computer? Think of the exciting worlds you could explore together.

Do ALIENS have PETS?

Have you ever had a pet? If you choose the right one, they can be the best friend you'll ever have. They give you love and attention, are never mean to you, and they don't ask if you've done your homework, or tidied your room.

I especially love dogs. They are loving, loyal and trustworthy. They would do anything to play ball or cuddle up with you on the couch. Cats are cool too! They're elegant, clean, friendly and purr happily on your lap when you stroke them. Rabbits? Well, they're just silly and cute.

Pets have been in our lives for a very long time. Around 20,000 years ago, wolves were first attracted to campfires by the smell of delicious food, cooking on the flames. The bravest and friendliest dogs thrived, wolfing down food scraps and sticking around at night to guard the camp. Eventually, dogs became tame enough to get regular pats and belly rubs. Everyone was happy with this arrangement.

Cats were domesticated at least 8000 years ago. They were popular pets in ancient times, with some even found mummified with Egyptian rulers in the pyramids. They became especially useful as they hunted the rats and mice that were eating precious winter stores of grain. Nowadays, pet cats are more likely to be found sunning themselves in a window, or curled up on a bed, asleep.

In the wild, animals and birds of different species are often found making friends.

On the African plains, you'll find ostriches and zebras hanging out. Ostriches have great eyesight and a long neck, and zebras have a fantastic sense of hearing and smell. Pairing up helps this odd couple sense danger when lions are prowling nearby. I wonder what they chat about when the coast is clear.

Cattle egrets are large white birds often seen perched on the back of cows in warmer parts of the world. They're not going for a joyride though, they are actually scouting for insects; including annoying ticks and fleas that they pick off the cow's ears and body. The egrets get an easy meal and the cows don't seem to mind giving their buddies a ride in exchange for a good clean!

Crocodiles have an unlikely friendship with a brave little bird called a plover. When the toothy beast sits in the sunshine with her mouth wide open, the bird darts in and picks out all the meaty goodness from between the croc's giant gnashers. It's a great feed for the bird, and a free dental clean and floss for the croc.

In animal sanctuaries and rescue centres, many other creatures (including monkeys, apes, dogs, ducks, goats, elephants and deer) form close friendships with other species. They sometimes even start acting like each other. Have you seen the video of the rescued baby rhino frolicking and leaping in the air with its lamb friend? You must, it's totally adorable.

So, if aliens exist, do they have pets? I don't see why not.

If we discovered enough planets that hosted life, we might eventually find a pterodactyl-like creature riding on the back of a magnificent stegosaurus-like beast, picking off fleas. Maybe there is a world out there with intelligent 6-legged insects out walking their pet spiders. Or dolphin-like creatures playing board games with their octopus-like friends. Perhaps giant blobs from the planet Miri have lesser spotted bumble bees as their companions, playing 'fetch' with the petals of the flowers that fall from giant wimble trees in autumn.

What creatures would superintelligent alien species keep as pets? Would they be real animals, or artificial intelligence avatars like characters in a computer game? The possibilities are endless.

Finally, would aliens take their pets on long space missions with them? Wouldn't it be great if the first alien that landed on planet Earth scuttled down the steps of their spacecraft accompanied by a 'labra-borg' (half-dog, half robot) in a space suit? As far as I'm concerned, any species intelligent enough to invent a dog toilet for a spacecraft deserves to have their best friend come along for the ride.

One DAY we'll MEET

On this incredible journey, we have met the many creatures of Earth, from dinosaurs to 11,000-year-old sponges. We've said 'howdy' to the mighty microbes that survive deep in ocean trenches, bubbling away in hot volcanic springs, floating high in the clouds, shivering on Antarctic rocks and donning space suits and blasting off to the Moon (okay, that's just the tardigrades).

Thousands of planets have been discovered beyond our solar system, some hot and molten, others frozen and gaseous. Many more will be found in future years, and some may be similar to Earth. But what we haven't (yet) discovered out there is extraterrestrial life.

No Morse code tapping out on our radios. No crackly video calls from lifeforms on Proxima b. And definitely no UFOs landing on the lawns of our leaders, bringing well-dressed ambassadors from the planet Zorg.

Why haven't we found alien buddies yet?

The most likely reason is that space is so very big. If aliens exist, we just haven't found them yet. With several light-years between neighbouring solar systems, any signals sent to Earth from aliens would be diluted by, or potentially missed in, the vastness of space.

Another reason we probably haven't found any other life out there is that we haven't been searching for very long. It's been only 90 years since we invented radio telescopes. Aliens could've been trying to call us for ages, whether for 400 years or 4 billion years. But unless their radio signals arrived in the past 90 years, we simply wouldn't have had the equipment to pick those messages up.

Maybe alien spacecraft have visited Earth in the past, but we missed them. An alien space probe could have flown by a billion years ago, scanning our planet for signs of intelligent life. What would they have found? Microbes. They would have crossed our planet off their list and returned to the inky blackness of space.

But hope is not lost. We're getting ever better at looking for alien pen pals.

Future spacecraft, like the James Webb Space Telescope (which will orbit the Sun after its launch in 2021), will hunt for signs of biological life in the atmospheres of thousands of nearby worlds. As a distant exoplanet passes in front of its star, scientists will measure the chemical signatures from its atmosphere as the starlight passes through. If chemicals like oxygen (created by photosynthesis) or methane (made by things like bacteria and released in cow farts) are present, the hunt will be on for other evidence of life including vegetation, radio communications or even starlight reflected from alien solar panels!

The next generation of radio telescopes, like the Square Kilometre Array, will sweep the skies for alien messages from across the cosmos. It will scan thousands of nearby planets for Earth-like radio communications and billions of distant planets for bright radio beacons from advanced beings. Our supercomputers will crunch trillions of numbers per second flowing from these telescopes to automatically scour millions of stars and galaxies for signs of extraterrestrial intelligence.

Is it important that we search for life?

Yes! But it's not just about finding new friends. Finding life 'out there' will help us understand more about life down here, on Earth. It might raise more questions than it answers, but being curious about how life in the universe began is a fundamental part of all human cultures. If science and technology can help us to keep one eye (and ear!) open for cosmic chatter, we owe it to ourselves to reach out to our celestial neighbours.

And, as we learn more about other worlds, we gain a greater appreciation of how rare and precious our home is. We must protect its delicate balance for the sake of the trillions of microbes, bumblebees, axolotls, dogs, cats and everyone you love, on this wonderful blue planet, Earth.

FURTHER EXPLORATION

It's not just professional scientists who get to look for other worlds, you can do it too! All you need is a computer, and a connection to the internet.

Astronomers use gigantic telescopes to study distant stars and look for clues as to whether they might have a family of planets orbiting around them. One way we hunt for planets is to measure how brightly the star is shining and wait for changes to the light that indicate that a planet is passing in front. This is like a mini eclipse.

Another way to find planets around other stars is to look for a 'wobble' in the position of a star, caused by the gravitational pull of planets as they move around in their orbits.

We can teach computers to scan the light output of stars and hunt for little dips in brightness, or to look out for star wobbles. But it turns out that human brains are actually often better at doing this than computers are!

That's why we can't rely on computers alone. We need your help. Modern telescopes study so many millions of stars and galaxies, we astronomers can't look through all the images ourselves. We need a community of helpers – called citizen scientists – to roll their sleeves up and get to work.

Wanna find your very own planets in solar systems throughout the Milky Way? Visit this website to get started today! exoplanets.nasa.gov/citizen-science/

ACKNOWLEDGEMENTS

My sincere thanks go to the editorial and creative team, Sally Heath, Sam Palfreyman, Rebecca Lim, Phil Campbell and Tracie Grimwood, for helping me create a volume that I'll be proud to see clutched in the eager hands of young space explorers around the world.

I'd also like to acknowledge Dr Erica Barlow, who lent her astro-biology prowess to the manuscript and improved my understanding of several key topics around the emergence of life on Earth.

Lisa Harvey-Smith is an award-winning astrophysicist and a professor at the University of New South Wales. With research interests in the birth and death of stars and supermassive black holes, Lisa also serves on the Australian Space Agency's Advisory Group. She previously worked on developing the Square Kilometre Array – a continent-spanning next-generation radio telescope that will survey billions of years of cosmic history.

Lisa has a talent for making complicated science seem simple and fun and is a presenter on the popular television show *Stargazing Live*, a guest on BBC's *The Sky at Night* and *The Infinite Monkey Cage*, and a regular science commentator on TV and radio.

Lisa has written 3 popular science books: *When Galaxies Collide*, *The Secret Life of Stars* and the bestselling children's book *Under the Stars: Astrophysics for Bedtime*. She has performed extensively in theatres with her self-penned *When Galaxies Collide* show and has appeared alongside Apollo-missions astronauts Buzz Aldrin, Charlie Duke and Gene Cernan.

As Australia's Women in STEM Ambassador, Lisa is responsible for increasing the participation of women and girls in Science, Technology, Engineering and Mathematics (STEM) studies and careers across Australia. She is also a vocal advocate for building inclusive workplaces for LGBTQI+ scientists. In her spare time, Lisa runs ultra-marathons, including multi-day, 12-hour and 24-hour races. She once ran 250 kilometres across Australia's Simpson desert.

To find out more, visit lisaharveysmith.com